BST

DISCARD

BST

Books by Fred Hoyle

The Black Cloud
Frontiers of Astronomy
Galaxies, Nuclei and Quasars
The Nature of the Universe
October the First Is Too Late

Books by Fred and Geoffrey Hoyle

The Molecule Men
Rockets in Ursa Major
Seven Steps to the Sun

NICOLAUS
COPERNICUS

AN ESSAY ON HIS LIFE AND WORK

BY

FRED HOYLE

HARPER & ROW, PUBLISHERS

New York, Evanston, San Francisco, London

1817

FIRST U.S. EDITION

Designed by Sidney Feinberg

Library of Congress Cataloging in Publication Data

Hoyle, Fred.
  Nicolaus Copernicus: an essay on his life and work

  Bibliography: p. 89
  1. Copernicus, Nicolaus, 1473-1543.
QB36.C8H75   1973   520'.92'4   [B]   73-4092
ISBN 0-06-011971-3

NAt

CONTENTS

Preface, ix

Contents

ILLUSTRATIONS

PREFACE

It is unexpectedly awkward
to describe exactly what Copernicus did, harder than to
write an account of the achievements of Kepler, just
because it is less difficult to write about the complete
answer to a problem than about an approximate one.
Kepler's solution to the problem of the planetary motions was essentially complete, while that of Copernicus
was approximate. To understand Copernicus we need
therefore to consider how good his approximations really
were, and this kind of question always turns out to be
difficult.

The difficulty is compounded by the style of the
work. Mathematicians and astronomers to the time of
Newton used intricate geometrical constructions, which
today we find unfamiliar. We prefer nowadays to use
symbols and equations instead of pictorial representations, because analytical methods have turned out to be
much more powerful—results are derived with economy

of effort, and are compact and therefore more easily understood. The proof that the planets move in elliptic orbits is a good example. Through the use of modern methods this result can be obtained from only a page of calculations and is readily comprehended by university students in a first course on mechanics. In contrast, Newton's original proof was probably not understood in his own day by more than a dozen people.

When I first encountered the geometrical representations of Copernicus and Ptolemy, I wondered how they could be described analytically. The answer turned out to be unexpectedly elegant. It became possible to interpret the many curious features of the geometry which they used. So on this five-hundredth anniversary of the birth of Copernicus I decided to write an account of the technical meaning of his work. This intention is covered in the present book by Chapters I, III, IV, and the Epilogue.

It appeared worthwhile to add the biographical sketch of Chapter II. In the outcome this sketch has caused me some difficulty. My idea was to abstract from standard biographies the aspects of the life of Copernicus which seemed relevant to his astronomical achievements. After consulting such accounts, of which the three volumes by Leopold Prowe (Weidmannsche Buchhandlung, Berlin, 1883–84) are the most complete, I found myself unable to answer certain simple and pertinent questions, particularly questions concerning the periods when the actual astronomical discoveries were made. I also found aspects of these accounts which I felt to be implausible, at any rate from the point of view of the working scientist. This inevitably led me into issues of interpretation, which I would much have preferred to avoid. For example, was *Commentariolus* written pre-1512 or post-1530? Above all, I became overwhelmed by the fact, discovered in the nineteenth century, that *De revolutionibus* was not set by the printer from Copernicus' own manuscript. Why? A fair copy must have been made by someone. Who? These questions contain implications which I have discussed toward the end of Chapter II.

It is once again my pleasure to thank Jan Rasmussen and Evaline Gibbs for their very considerable help in the preparation of the manuscript of this book.

1973   F.H.

I

PRELIMINARIES

THE science of the ancient Mediterranean civilizations was weak compared to modern science because it contained no real understanding of dynamics. In retrospect we see that two stumbling blocks were responsible for this shortcoming. One lay in the concepts of "mass" and "force," as being determined through departures from uniform motion. The other lay in the mode of description of the planetary motions. Let it be understood at the outset that it makes no difference, from the point of view of describing planetary motions, whether we take the Earth or the Sun as the center of the solar system. Since the issue is one of relative motion only, there are infinitely many exactly equivalent descriptions referred to different centers— in principle any point will do, the Moon, Jupiter. . . . So the passions loosed on the world by the publication of Copernicus' book, *De revolutionibus orbium caelestium libri VI,* were logically irrelevant, although the controversies of the Counter Reformation served an important practical purpose, as we shall see later in this chapter. While this matter of relative motion is clear enough to

Figure 1. North-Central Europe, showing the province of Warmia at the time of Copernicus.

the modern mind, it was not at all clear in the sixteenth century. Tycho Brahe proposed a dualistic scheme, with the Sun going around the Earth but with all other planets then going around the Sun, and in making this proposal he believed he was offering something radically different from Copernicus. And in rejecting Tycho's scheme, Kepler obviously thought so too. Yet in principle there is no difference.

So what was the issue? The issue was to obtain even *one* substantially correct empirical description of the planetary motions. The issue was to find out *how* the planets moved. Not until this problem was solved could the more far-reaching problem of finding out *why* be tackled effectively. The essential facts had to come before the dynamical theory could be developed. The modern student learns the answer to the problem of *how* in one quick gulp. Each planet moves essentially in an ellipse with the Sun at one of the foci. To describe the actual motions one needs to know the orbital period of each planet, the size and eccentricity of its ellipse, and the orientation and the position of each planet at some instant of time. With knowledgeable hindsight, the situation may not seem unduly complicated, but looked at without foreknowledge the problem of *how* is anything but simple. This can be seen from Plate I, which simulates the bare facts of how a terrestrial observer sees the planets move over a period of time. If one started with observations of this kind, it was anything but easy to discover the rule of elliptic motion. I doubt whether it could subjectively have seemed less difficult than discovering the rules of structure in modern particle physics. In both these cases, separated in our science by many centuries, the procedure is the same—first find the empirical facts, then find systematic regularities, then find the reason for the regularities.

A quick answer to the question of why the work of

PLATE I. The motions of the planets over a period of some twenty years have been simulated to show how they appear to an observer on the Earth (Munich Planetarium).

Copernicus was so important would be to say that by placing the Sun at the center of the solar system he made it easier to arrive at a substantially correct description of planetary motion—that it proved more straightforward to grapple with the problem that way. I think this answer contains a measure of truth. Certainly it was easier a century later for Newton and his contemporaries to tackle the dynamical problem in terms of a heliocentric description. But such an answer overlooks the pre-Copernican situation, in which the geometric theory of Ptolemy had proved more successful than the heliocentric theory of Aristarchus. Until Copernicus, experience was just the other way round. Indeed, Copernicus had to struggle long and hard over many years before he equaled Ptolemy, and in the end the Copernican theory did not greatly surpass that of Ptolemy. I say this not to diminish the achievement of Copernicus but in the hope that the five-hundredth anniversary of his birth will prove an opportunity for the modern world to pay tribute also to the greatness of Ptolemy.

Let us see in general terms what was achieved. To have any hope of representing the motion of a planet in an approximate but controlled way it is necessary to know the synodic period to high accuracy.* By a controlled way I mean one free from cumulative errors, errors that start by being small but which pile up in-

* The synodic period is the period of revolution as seen from the Earth.

exorably to larger and larger values as time goes on. So long as we arrange our theory so that each planet completes its orbit in the sky in the correct period there will be no such progressive errors. The errors will be of detail, resulting from our inaccuracies in describing the orbit itself. To obtain substantially the correct periods of the planets was easy, at any rate for the whole human species if not to an individual astronomer. All that need be done is to observe a planet over a large number of revolutions. We judge a revolution to be completed when the planet returns to the same position with respect to the stars. Our observation in this respect need not be very accurate provided sufficiently many revolutions are observed. Suppose we can judge position with respect to the stars to within $0.5°$, using simple naked-eye observations. This is about $1/700$ of a revolution. So if we observed just one revolution we could expect to mistake the orbital period by about one part in 700. But after n periods we shall determine the period to within one part in $700\,n$, which for n large becomes very accurate. Now the planets have been observed from this point of view from time immemorial, and good values for their periods were available to all makers of theories throughout historic times, to the Greeks as well as to Copernicus.

With this in mind we see that a lack of knowledge of the precise orbit, and of the dynamical behavior in the orbit, can only result in detailed errors. At any given moment a planet will be observed to be out of its theoretical position by some moderate angle, the amount depending on the quality of the theory. It is interesting to classify approximate theories according to their relation to the eccentricity of the actual elliptic orbit. In Figure 2 we have an ellipse representing a planetary

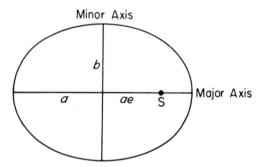

Figure 2. The planetary orbits are ellipses, being less flattened than the ellipse shown here. Even when the ellipse flattens only slightly, the focus S moves markedly away from the center. It is this property which caused glaring discrepancies to appear in approximate theories, particularly when applied to the planet Mars.

orbit, with semi-major axis a, semi-minor axis b, and eccentricity e. The Sun is considered to be at the focus S. It is the nature of an ellipse that the lengths of the axes are related to the eccentricity by

$$\frac{b^2}{a^2} = 1 - e^2 \,. \tag{1}$$

In the particular case $e = 0$ the axes are equal, $a = b$, and we have a circle. As e increases, approaching unity, the ellipse becomes flatter and flatter. No cases with e approaching unity are found among the planets, although such cases occur often for comets. The situation for the six planets known to Copernicus and to the ancient world is shown in Table 1.

TABLE 1

Planet	Eccentricity	Major axis (in terms of Earth)
Mercury	0.2056	0.387
Venus	0.0068	0.723
Earth	0.0167	1.000
Mars	0.0933	1.524
Jupiter	0.0484	5.203
Saturn	0.0558	9.539

In all cases the orbits are nearly circles. Even for Mercury the shorter axis is only about 2 percent less than the larger axis. This was a most fortunate circumstance since it permitted approximate theories to be developed that were based on circles.

If one makes the approximation of replacing the elliptic orbit by a single circle with center at the Sun, the angular errors between the observed positions of a planet and the theoretical positions will be of the order of the eccentricity. This is in terms of the unit of angle,

which is a fraction $\dfrac{1}{2\pi}$ of a full revolution, about 57°.
For Mercury, the planet in our table of largest eccentricity, the errors as seen from the Sun turn out to average about 15°. As seen from the Earth, the errors will
be less than this, however, because Mercury is nearer
to the Sun than it is to the Earth. Seen from the Earth
the errors will average more like 5°. For Mars, the planet
in our table with the next largest eccentricity, the errors
average about 7° seen from the Sun. But at opposition
Mars approaches much closer to the Earth than the
Martian distance from the Sun, and this greatly exag-
gerates the error for a terrestrial observer, to well over
10°. Thus for an observer on the Earth errors will be
largest for Mars. It is curious that although the actual
orbits do not differ in shape much from circles the errors
of a circular model can nevertheless be quite large.

Indeed, errors as large as this were quite unacceptable
to Greek astronomers of the stature of Hipparchus and
Ptolemy. It was this, rather than prejudice, which
caused them to reject the simple heliocentric theory of
Aristarchus. Little is known of the circumstances in
which Aristarchus put forward his theory. There are
indications that the approach may have been somewhat
tentative, which may well have been so in view of his
likely knowledge of the resulting discrepancies between
the theory and observation.

The planet Mars has played a great role in the history of astronomy. Quite apart from its importance in the Aristarchus–Ptolemy–Copernicus–Kepler story, it served to cast serious doubt on an earlier theory, the homocentric spheres of Eudoxus. It is worth bringing this point to mind because it has a relation to the humanistic intellectual activity that was sweeping through Europe at the time of Copernicus. All the spheres of Eudoxus had the Earth as their center. The outermost sphere had the stars attached to it, being otherwise dark. This outermost sphere rotated to give the diurnal motion of the heavens. Inside this sphere were many other, translucent ones. The Moon and the Sun were given nests of three each, while the planets each had a nest of four spheres. The outer sphere of a nest had the same motion as the sphere of the stars. The second outermost sphere of each nest was attached to the outermost by a polar axis about which it was free to turn. The third one was attached to the second in a similar way, and so on. Finally, the planet, or the Sun or Moon, was attached to the innermost sphere of the nest. In this way complex motions of the various bodies could be produced. The mathematical problem was to choose the polar axes and their points of attachment and the motions of the spheres in such a way as to reproduce the observed behavior of the Sun, Moon, and planets. It will be clear that this theory contained so many parameters that a fair measure of success for it could be expected. But it failed in a critical respect. Because the spheres were all centered at the Earth, the motions had perforce to be of such a kind that any particular planet, or the Sun or Moon, remained always at the same distance from the Earth. Hence each planet should maintain a constant brightness—which is not true. Particularly for Mars there are large variations of brightness, because the distance of Mars changes markedly, being some five times greater at solar conjunction than at opposition.

Even so, the theory of Eudoxus found great favor with Aristotle, who extended it himself until it came to involve more than fifty geocentric spheres. This was the theory that came to the notice of the European humanists during the fifteenth century. Access to Greek culture, accorded the same respect as the teachings of Aristotle, contrived to draw an enormous red herring across the path of the serious astronomer. Copernicus, in particular, had to fight a war on two fronts. On the one hand he had to do fierce battle with the theory of Ptolemy—which was of an altogether higher level of sophistication than the theory of Aristotle—and at the same time he had to combat the prejudice toward a geocentric theory which was being so strongly supported by the popular adherence to Aristotle. The theory of Ptolemy, a few minor imperfections apart, worked correctly to the first order in explaining the planetary eccentricities. Copernicus with his heliocentric theory had to do at least as well as this, which meant that *he had to produce something much better than the simple heliocentric picture of Aristarchus.* Since he was successful in this, the outcome was the heliocentric theory published in *De revolutionibus.*

In Chapter IV we shall find that, judged in terms of predictive capacity, there was little to choose between the new heliocentric theory and the theory of Ptolemy. The new theory was crucially important, however, just because of the intense emotional reaction to it. Emotional opposition was so strong that astronomers and astronomy came very much into the popular eye, whether favorably or adversely is not relevant. The essentials were that astronomers became impelled to observe the planets with increasing accuracy, and funds to enable them to do so were forthcoming.

The key issue lay, once again, in the motion of Mars. By working correctly to the first order in the eccentricity, the discrepancies between theory and observation were reduced to terms of the order of the square of the eccentricity. But this gave a maximum error for Mars of more than $1°$, which was still very glaring. Persistent discrepancies for Mars, taken with the highly charged controversial situation, forced an attempt at a more far-reaching theory. The time became ripe for increased understanding.

The Copernican theory fitted the planets of small eccentricity—Venus, Earth, Jupiter, and Saturn—very well, to within the errors of observation in some cases, so there had to be something "right" about it. The theory thus provided a base from which to proceed. Progress was certain as soon as a man, or men, of adequate stature came along. As we know in retrospect, favorable circumstances came with the observations of Tycho Brahe, and with the appearance of Kepler. Kepler fitted the task of advancing the theory, not only in stature, but in the demonic energy with which he tackled the problem of Mars. The sheer volume of calculation which he carried through fits him to be called the iron man of science—much as Richard Wagner may be called the iron man of music.

In the beginning Kepler contented himself with setting right one or two minor blemishes in the Copernican theory. Finding this did little to help the problem of Mars, he then began a long series of trial calculations, which in effect were a search, along Copernican lines, for the missing second-order terms in the planetary eccentricities. Kepler achieved improvements, but not complete success, and always at the expense of increasing complexity.

Kepler and his successors might well have gone on in this style for generations without arriving at a satisfactory final solution, for a reason we now understand clearly. There is no simple mathematical expression for the way in which the direction of a planet—its heliocentric longitude—changes with time. Even today we must express the longitude as an infinite series of terms when we use time as the free variable. What Ptolemy, Copernicus, and Kepler, in his early long calculations, were trying to do was to discover by trial and error the terms of this series. Since the terms become more complicated as one goes to higher orders in the eccentricity, the task became successively harder and harder. The modern expression for the heliocentric longitude θ measured from perihelion in terms of the time t is

$$\theta = nt + 2e \sin nt + \frac{5}{4} e^2 \sin 2nt$$

$$- e^3 \left[\frac{1}{4} \sin nt - \frac{13}{12} \sin 3nt \right] + \ldots \quad (2)$$

Here e is the eccentricity and n is the planet's "mean motion," defined in terms of the sidereal period P by

$$n = \frac{2\pi}{P} \,. \quad (3)$$

The heliocentric distance r is related to time by

$$\frac{r}{a} = 1 - e \cos nt + \frac{1}{2} e^2 (1 - \cos 2nt)$$

$$+ \frac{3}{8} e^3 (\cos nt - \cos 3nt) + \ldots (4)$$

The forms of the terms in these series are not easy to guess, and if guessed would not seem at all natural or elegant. Notice that one has to guess *both* series in order to be able to relate predicted heliocentric values to the observed geocentric longitudes of a planet. As well as knowing P one must also know a and e, although the semi-major axis a need not be known in the usual local units (meters). It is sufficient to know the values of a in relation to one another, for example in relation to a for the Earth. Copernicus had tolerable values of P, reasonably good values of e, and quite good values of a.

TABLE 2

Planet	Copernican value of a	Modern value
Mercury	0.3763	0.3871
Venus	0.7193	0.7233
Earth	1.0000 (standard)	1.0000
Mars	1.5198	1.5237
Jupiter	5.2192	5.2028
Saturn	9.1743	9.5388

To come back to Kepler, the stroke which permitted Kepler to solve the problem came from abandoning a frontal attack on the problem of guessing the two series given above. Instead, Kepler set himself the problem of determining the *shape* of the orbit, which is to say determining r, not in terms of time, but in terms of θ (cf. Figure 3). With this approach the answer was elegant and reasonably straightforward: r varied with θ like

$$\frac{1}{1 + e \cos \theta}. \tag{5}$$

Kepler knew immediately that this relationship defined an ellipse, and that the constant of proportionality between r and equation (5) had to be related to a and e in the usual mathematical way, namely:

$$r = \frac{a\,(1 - e^2)}{1 + e \cos \theta}\,.\qquad(6)$$

There remained the problem of finding how a planet moved in a *now known* orbit. This problem was solved by another remarkably elegant discovery: a planet

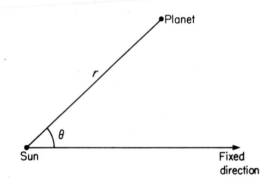

Figure 3. The polar coordinates r, θ of a planet taken with respect to the Sun. The fixed direction here is to be interpreted as fixed with respect to distant parts of the universe (see Epilogue).

sweeps equal areas of its ellipse in equal times, the time for the whole ellipse being just the known period P.

Another remarkable discovery was that the empirically determined values of P and a were such that P^2 was proportional to a^3, a fact which could have been known to Copernicus but which does not seem to have been noted by him. This relationship remained empirical, as did Kepler's other discoveries, until a dynamical theory became available eighty years later.

So quite suddenly the apparently intractable problem of describing the planetary motions was made to look simple. From the moment Kepler found equation (6) the course of modern science was set. Until then there had been nothing really firm for science to get its teeth into. Wipe out all that has been discovered since Kepler. Go back to the year 1610 and do it all again. Let there be no Newton this time, no Euler, no Lagrange. . . . Yet there will be others making the same discoveries, perhaps more slowly but probably not much more slowly. The flowers along the way may be different, but the path will be the same. It is because Copernicus focused the attention of the world at precisely the right spot, the place where Nature simply had to give up her secrets, that today we judge his work to have been so important. In mountaineering jargon, Copernicus found the "point of attack." Let us now go back to his own times, and see how he lived and worked, and what manner of man he seems to have been.

II

MR. STANDFAST

The name as we know it, Nicolaus Copernicus, is a Latinized self-name, adapted for scholarly writings from the family name of Niklas Koppernigk. Copernicus, as I shall continue to call him, signed himself "Coppernic" on official documents, closer to the Koppernigk form. The family seems to have migrated from Silesia, moving into Poland early in the fifteenth century. It was commonplace for German-speaking settlers to move into Poland throughout the fourteenth and fifteenth centuries. Indeed, the Teutonic Knights, an order founded during the Crusades, had already been invited into Poland in the thirteenth century, to help, it is said, with certain territorial ambitions of the Polish rulers. For their services the Knights were given the eastern Baltic region which we know today as East Prussia. Between East Prussia and Poland lay the little province of Warmia, destined to play a critical role in the life of Copernicus.

With its northern flank reasonably secure, Poland in

the latter part of the fifteenth century was pressing outward over a wide arc, from present-day Czechoslovakia and Hungary in the southwest to Russia in the east. Poland controlled a long wide corridor stretching south from the Baltic almost to the Black Sea.

Such a broad pattern of life could hardly have failed to have influence on Copernicus. To grow up in an era of free movement of peoples is an advantage toward

PLATE II. Torun, where Copernicus was born on February 19, 1473.

acquiring an unprejudiced mind, always assuming the times are not so violently disturbed as to make sheer physical survival a major demand. Copernicus was doubly lucky here. He acquired an independence of outlook while enjoying the security of a stable family. One Johann Koppernigk, appearing in the records of Cracow as merchant and banker, was probably the father of Niklas Koppernigk, also merchant and banker. Niklas moved from Cracow to Torun (Plate II) on the Vistula, and it was here that the second Niklas Koppernigk, the astronomer, was born on February 19, 1473. With both father and grandfather immersed in the field of commerce and banking, it should not surprise us to find Copernicus in later life writing a well-reasoned monograph on practical economics, which in the outcome had a significant effect on the money systems of Poland and East Prussia.

Niklas the elder died when the future astronomer was only ten. Thus it came about that Copernicus was adopted by his uncle, Lucas Watzenrode (Plate III), who six years later was to be appointed Prince-Bishop of Warmia. Since Watzenrode appears to have been a man of high temper it is something of a mystery how he came to be appointed to such a post. At all events he came to a position where he could be of great material help to Copernicus, who was never to know the crushing disability of economic privation.

Copernicus was soon traveling to Cracow, on his uncle's advice or instruction, to begin studies at the university there. He was fortunate in that Adalbert Brudzewski (1445–97), professor of astronomy at the university (although moving from astronomy to philosophy at about this time), had been a pupil of Johann Müller (Regiomontanus, 1436–76), who was himself the follower of Georg Peurbach (1423–61) at the University of Vienna. Thus Peurbach's book *Epitome in Ptolemaei Almagestum* probably became known to Copernicus at

PLATE III. Lucas Watzenrode (Watzelrode), uncle of
Copernicus, appointed Prince-Bishop of Warmia (*ca.* 1489).
The election was a snap affair, hurried through in less than
a week. The Chapter (Diocese) of Warmia was anxious to
secure autonomy in the appointment of its bishops. The
choice of Watzenrode was expected to be acceptable to the
Polish King. While this may have been true in a personal
sense, the manner of his election was to be a source of
contention throughout the rest of his life. The squabbles
over this issue involved Copernicus closely in the years from
1503 to 1512.

the very beginning of his career. During these early student days the university is said to have received a present of astronomical instruments, primitive indeed by our standards, but very up to date at the time. Probably Copernicus played a part in setting up these instruments and in making observations with them, for a few years later when taking up studies in Italy we find him working with Maria da Novara (1454–1504), professor of astronomy at Bologna, "not so much as a pupil as an assistant." This paved the way for the observational work which Copernicus was to need in his later years.

A truly momentous event occurred during these Cracow days. Copernicus was nineteen in the year Columbus discovered America. We have no account of the impact this discovery made on young Copernicus, but it needs little imagination to understand how it came about that he turned so wholeheartedly to the views of the Greek, Heraclides, who first suggested that the Earth is a sphere and that the apparent diurnal rotation of the heavens is due to rotation of the terrestrial sphere, not to an actual motion of the heavens themselves. It is indicative of how far medieval Europe had advanced beyond the Greeks in the understanding of practical physical problems that Copernicus, when confronted by the old argument that rotation would cause the Earth to fly apart, answered by remarking how much more surely would the heavens fly apart if called on to rotate at a similar rate. Here we have the sound physical perception that the forces developed for a given rate of rotation are greater in a large system than in a small one. Although *De revolutionibus* is concerned largely with abstract geometrical constructions, it is as well to remember that behind all this formal work lay a great deal of physical common sense. Indeed, common sense, or sound sense as we might more properly say, was the epitome of Copernicus' life.

In 1496 we find Copernicus traveling, again on his uncle's advice, from Poland to the University of Bologna. Together with his brother, Andreas, he is said to have walked across the Alps. It is a matter for regret that neither brother apparently saw fit to commit an account of the journey to paper—mountaineers as well as astronomers would richly prize such a journal.

In 1497, through Lucas Watzenrode's good offices, Copernicus was appointed a canon of the Chapter of Warmia. Lucas would seem to have found this nepotism not altogether easy, for he brought it off only on his second attempt. Copernicus now had a measure of financial independence which he was to retain for the rest of his life. Andreas had to wait two more years before receiving his canonry, but Andreas was not to profit from the appointment in the same degree as Niklas, for he was to die young—possibly from leprosy.

Copernicus returned to Warmia in 1501, to be inducted into his canonry. Then he was back in Italy, now at Padua studying medicine. This is usually thought to have been at his uncle's instigation. However, Copernicus' mother, Barbara, was ill and indeed nearing the end of her life at this time, which would seem to give a more plausible reason for the return to Poland—and more than one young man has been impelled into the study of medicine through the illness of his mother. Copernicus returned finally to Poland in 1503. His student days were truly over at last. Thirty years old now, he had studied mathematics and astronomy, medicine and law, he had a knowledge of many languages, latterly including Greek, and he was a master of most of them. This was the pattern of the Renaissance man. It is a pattern which many educators (the devil stir them) seek after in our own time. You were fortunate if you were born a man of many parts in the Renaissance, but the child born today with abilities in many fields is far less fortunate than it may seem at the outset. He will be encouraged by the educators to develop "broadly." Likely enough he will please his teachers by turning in good performances over a wide range of subjects, but he will end up with the sad discovery that there is no field in which he can compete successfully with the narrow specialists whom our educators deplore so loudly. The chances are that the widely gifted child will wind up today as an embittered person without real self-confidence. But it was otherwise during the Renaissance, and it was very much otherwise with Copernicus. He could afford the many years of student life because

PLATE IV. Copernicus, based on a self-portrait.

there were no narrow specialists treading on his heels. Copernicus even made a stab at the arts, but Plate IV, based on a self-portrait, shows him to have been no Leonardo da Vinci.

How far toward the *De revolutionibus* was Copernicus at the end of his student days? We do not know and can only speculate. At some time before *De revolutionibus* Copernicus sent out a manuscript, *Commentariolus*, in which he describes in a kind of extended abstract the ideas of his heliocentric theory, leaving aside for a later, larger work the more difficult mathematical demonstrations. Although certain of the constructional methods used in *Commentariolus* are different from those in *De revolutionibus*, it is clear that *Commentariolus* was written after Copernicus had arrived at the main structure of his theory. So if we knew the date of *Commentariolus* we should be well along the road toward answering our question, but one scholar has given a date post-1530 and another pre-1512, which covers quite a range of possibilities! If we accept the earlier date, much of the work must have been done before the return to Poland in 1503, since on his return Copernicus became secretary, legal adviser, and personal physician to his uncle. He resided until 1512 at Heilsberg (Plate V), where he became involved in many a sharp political situation. Lucas Watzenrode, a tornado of a man, was inclined to deal hotheadedly with political problems.

We find Copernicus calming the old fellow; we find him writing reports and memoranda, advising on negotiations, acting as physician—and in earnest too, since his uncle was not a fit man. We find him traveling, locally

PLATE V. Heilsberg.

and nationally. These activities are not of a kind to go well with the pondering of a deep intellectual problem. I believe, then, that if Copernicus, by 1509 or thereabouts, was in possession of the guts of the solution to the heliocentric problem, he must of necessity have had most of it already in 1503. In which case I find it surprising that he left no trail of curiosity and wonderment behind him in Italy.

In 1509 Copernicus published the first translation from Greek into Latin of a collection of ribald letters by the Byzantine poet Theophylactus Simocatta, the book being dedicated to Lucas Watzenrode. This translation does not fit with a man concentrating deeply on an abstract mathematical problem. If Copernicus were really immersed in his theory over the hectic period 1503–12, surely he would have snatched every spare moment for his astronomical investigations. He would hardly have had time to spare for Theophylactus. It is a curious thought that but for the almost accidental circumstances which seem to have led to the publication of *De revolutionibus* Copernicus would have gone down as the author of two books, one of ribald letters, the other a dissertation on money policy.

We should be grateful for the matter of these ribald letters, because this book of 1509, and an episode from Copernicus' later life, provide us with rare glimpses of the man in flesh and blood. Except in this one moment I cannot hear Copernicus laugh. I hear Kepler chuckling away throughout his life. I hear great gusts of laughter from Galileo. I get none from Newton, though—only sniffs—because I do not feel there was any laughter in Newton as a person. For Copernicus the situation was different. The portraits of the fifteenth century stare down at us from the walls in impassive masks. The fifteenth century was an age in which men learned to keep their true selves hidden from the world, especially men with the responsibilities now falling onto the shoulders of Copernicus.

To give my own appraisal of the astronomical situation, I suspect Copernicus arrived at his position concerning the rotation of the Earth in his university days. I suspect Copernicus, as a student, became acquainted with a publication, *An terra moveatur an quiescat,* in which Johann Müller (Regiomontanus) sets out with approval all the old arguments against the rotation of the Earth, in which he comments sarcastically on the idea of the Earth being roasted like meat on a rotating spit. If Copernicus had decided already in favor of the Earth's rotation, say, after Columbus in 1492, he would have felt cause to doubt all the other arguments of Regiomontanus. In the manner of the young, he would have been inclined to regard the old man as a bit of a fool, and he would have come to discount Müller's arguments in favor of the Ptolemaic theory. I see the young Copernicus with the conviction that one, at least, of the main ideas of the old theory was wrong, and if one of its cornerstones was gone perhaps the old theory was wrong as a whole?

After his uncle's death in 1512, Copernicus at last took up his canonry at Frauenburg, a town at the mouth of the Vistula (Plate VI). The cathedral there overlooks a fresh-water lagoon opening into the Gulf of Danzig. This was a drastic change, in all material and outward respects. Instead of his being in many ways the most important man in Warmia (the secretary of any

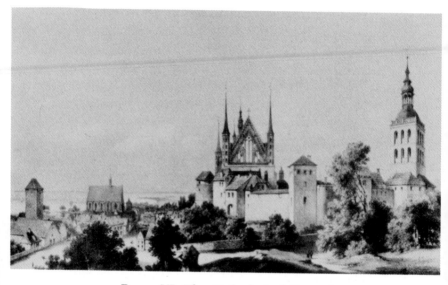

PLATE VI. The Cathedral of Frauenburg.

organization is usually its most important man, *de facto* if not in name), Copernicus was now just a canon among many canons (there were fifteen others) domiciled in a not particularly important town. There can be little doubt that such a change must have involved a considerable drop in what is nowadays called the "pecking order." Biographers have made much of the emergence of Copernicus in this period as a physician of widespread renown, eagerly sought after by both high-born and low. Such an attempt to represent his life as the unimpeded voyage of overriding genius seems questionable. Unless the men of his day were very different in emotional make-up from those of our own, officials at the headquarters of the Chapter of Warmia at Allenstein must have grown restive—to say the least—at the influence which Copernicus exercised over his uncle, Lucas Watzenrode. Officials never relish a situation in which a youngish man from outside comes to be promoted above their heads, as Copernicus was in the years from 1503 to 1512. In such circumstances it is not the way of officials to move openly; they lie low biding their time. When their time comes, as it did with Watzenrode's death, they step in to take revenge. It is in such a light that I interpret the removal of Copernicus to Frauenburg. And I interpret the medical activities of Copernicus as a reaction to demotion. By the respect accorded to him as a medical authority he contrived to maintain his reputation as a man of consequence.

To take this attempted interpretation a bit further, I would see something of an ambivalent attitude developing at Allenstein in respect to Copernicus. Immediately following Watzenrode's death I see the situation as I have just described it. But after a few years had slipped away, after immediate hostilities had weakened, it would be remembered how very able Copernicus was.

The authorities at Allenstein, like cats in a box, would have found new quarrels. There would perhaps be among them some who would find themselves contemplating the possibility of his return, not to a top-level job, but to something a bit humdrum requiring precise, competent administration. It is in these terms that I see the recall of Copernicus in 1516 to the job of administering certain outlying estates of the Chapter. This was a far cry from the big issues with which he had been concerned during his uncle's lifetime. It is of interest that he accepted such a post at all.

The storm clouds of war were gathering during the years Copernicus occupied this minor position. In 1520 the Teutonic Knights under Albrecht von Hohenzollern (1490–1568) attacked Poland across Warmia in an attempt to secure independence for East Prussia. One of the few positive aspects of such a situation is that men of ability rise rapidly. Petty animosities become forgotten in the urgency of the moment. So it was with Copernicus. We find him administering outlying estates no longer. By 1519 he was back essentially at his old job, as Secretary and Chancellor of the whole Chapter of Warmia. I see no reason to doubt the story that, when the Knights besieged the capital of Allenstein, senior officials fled the city, leaving Copernicus behind as Governor-in-Charge. It is consistent with all we know of his ability and character that the city did not fall.

Nor were the Knights successful in the war generally.

The outcome was devastation and inflation throughout Warmia. Copernicus was called on to prepare a memorandum on the state of the country, a document which eventually served the Polish King as a basis for peace negotiations. It was at this time that Copernicus became deeply concerned with problems of economics. He became involved with farmers, helping them rebuild and restock their herds. He sought to maintain fixed prices and to maintain standards for staple commodities. In 1523, in an interregnum between bishops, he acted as Administrator General of all church property in Warmia—a far cry indeed from the minor post of 1516.

The merchant-banking skills of the Koppernigk family came to the fore in this period. Copernicus wrote a tract, *Monetae cudendae ratio,* in which he proposed currency reforms, some of immediate good sense, others far-reaching into our modern world. Up to this time, different cities in Poland and Prussia had each minted their own coinage. This produced a situation analogous to that which occurred in modern Germany after the last war. (The Allies agreed at Yalta to use a common currency. While this was sensible enough, neither Roosevelt nor Churchill insisted on defining how much currency each power should print. The inevitable result of this omission was that the Russians swamped the market, inflating money to whatever degree was necessary to prevent German recovery.) To allow normal trade and interchange between cities there had of necesrode, but now Watzenrode was also dead. So, too, was Copernicus' mother. By this time, Andreas, his companion in the walk across the Alps, was sick with an incurable ailment. Of his sisters, Barbara had retreated from the world into a nunnery. Only Katherina, married to a Cracow merchant, effectively remained of his family. Copernicus in 1513 must have felt very much alone. Only ten years earlier he had returned from carefree student days in Italy. Now, after a decade of hectic activity, the world had suddenly taken on a darker hue.

It appears likely to me that the year 1513 marks a decisive turning point in the life of Copernicus. It is to the period 1513–16 that I would place the first work in astronomy *at a fully serious level.* I suspect this may well have been the first time in his life that Copernicus set himself to meet the full quantitative challenge of the Ptolemaic theory. To this point, I would expect him to have developed a clear qualitative insight into the planetary problem, but I think the guts of the Copernican theory as we know it still remained to be worked out. I see the situation much as I see the deafness of Beethoven. I do not believe Beethoven would have developed the full measure of his greatness had he not become deaf. I similarly doubt if Copernicus would have driven through the more difficult aspects of his theory had it not been for the adversities which befell him in the year 1513.

It is worth considering objections to this point of

view. Why in this case did Copernicus accept the minor administrative post in 1516? There are several good answers to this question. First, as a matter of self-respect, Copernicus must have wanted to prove himself in the world of affairs, to prove that he could succeed without the patronage of his uncle. Second, the post, being a minor one, would not interfere too seriously with astronomical work. Third, the storm clouds of war were gathering. Copernicus may have wanted to "do his bit," as many scientists did in the last war. Certainly the increasingly heavy involvement over the years 1519–23 can be explained in these terms. It is not unreasonable to suppose that the war prevented Copernicus from following his intentions until after 1523. Then, one can fairly argue, he felt able to return to his personal researches—and this was why he elected to return to Frauenburg, instead of continuing at Allenstein. It is to the following decade, 1523–33, that I would attribute the quantitative core of the Copernican theory.*

In 1513, Copernicus was already forty. There may be critics, mindful of the modern adage which says that the world's best theoretical physicists are all worked out at thirty, who will consider forty too ad-

---

* From a recent analysis of the watermarks of the paper actually used by Copernicus, it has been inferred that the first pages of *De revolutionibus* were written in 1519 and that the last pages date from about 1538. Amendations of the text may have been made even later than 1538.

vanced an age for Copernicus to have begun an assault on the most crucial aspects of his theory. To this criticism I would reply that if modern physicists are worked out at thirty, it is because they have grossly overworked in earlier years, and that the condition of being "worked out" consists of boredom and lack of energy, not of lack of ability. Newton (1643–1727) wrote his *Principia* when he was about forty. Schrödinger was approaching forty when he found his wave equation. Beethoven was in his fifties when he wrote what many feel to be his finest music, and Verdi was twice the age of forty when he wrote the tremendous trumpet passage at the end of *Otello*.

Let us proceed then on the basis that Copernicus arrived at the qualitative aspects of his theory in the period up to 1503, that he began serious quantitative work only in 1513, and that the full measure of the theory finally took shape in the decade from 1523. This places 1533 as a likely approximation to the date of *Commentariolus*, a dating which fits the known facts in a reasonable way. It may well have been for Johann Widmannstadt, secretary to Pope Clement VII, that *Commentariolus* was written. Widmannstadt is known to have presented Copernicus' ideas to Pope Clement in 1533. In 1536, as an outcome of discussions within the Church, Copernicus received a letter from Cardinal Schönberg urging that a full account of the theory be published (Galileo rest in peace). Word of the new theory also reached Protestant circles. In the spring of 1539 Copernicus received a visit from Georg Joachim von Lauchen (Rheticus). In Wittenberg, von Lauchen had heard a synopsis of the theory. His interest being aroused, he decided to learn more of it from Copernicus himself. Intending at first to spend but a few weeks or months in Frauenburg, von Lauchen in fact spent two years there. It is usually said that von Lauchen occupied himself over this extended visit in familiarizing himself with the detailed manuscript which was eventually to be published as *De revolutionibus*.

In the light of these facts it is hard to accept an early date for *Commentariolus*. It scarcely seems possible for Copernicus to have sent out a quite substantial account of his theory as early as 1512 and for it to have produced no significant reaction until 1533.

The circumstances in which *De revolutionibus* came to be published have often been discussed. It is generally accepted that Copernicus was exceedingly reluctant to publish. Von Lauchen is usually credited with persuading him to do so. Yet high dignitaries of the Church were suggesting publication long before von Lauchen appeared in Frauenburg—the letter from Cardinal Schönberg was dated November, 1536—and Tiedemann Giese, a close and valued friend, was also firmly urging publication. It would seem, then, that the factors which deterred Copernicus must have been either scientific or personal. Possibly von Lauchen played a part, which neither Schönberg nor Giese could play, of allaying certain scientific misgivings. I will return to this interesting possibility after considering the more personal issues.

Copernicus wrote in July, 1540, to Andreas Osiander asking if it would be possible for the theory to be published without exciting hostile criticism. Osiander replied in the spring of the following year: "For my part, I have always felt about hypotheses that they are not articles of faith, but bases of calculation, so that, even if false, it does not matter so long as they exactly represent the phenomena of the celestial motions. . . ." In this interchange one detects a fear of criticism. On another occasion Copernicus remarked quite explicitly that he was "afraid of scorn and contempt on account of the novelty and inconceivable nature of my theory." But scorn and contempt from whom?

A new Prince-Bishop of Warmia, Jan Flachsbinder (Joannes Danticus), was elected in 1537. It seems likely that Copernicus played a considerable role in this election, opposing the new Bishop in favor of another candidate—possibly Tiedemann Giese. Copernicus then became subject to petty persecution, and to a particular prohibition which appears to have affected him deeply. Anna Schilling, daughter of a cousin of Copernicus' mother, a widow and a woman of considerable culture, had come to Frauenburg to act as his housekeeper. On moral grounds, the new Bishop insisted she must leave. Copernicus fought the case to the limits of what was possible, but his petition to retain Anna was not successful. We see vividly in this issue the extent to which Copernicus was not his own master in matters we today would take for granted. Copernicus was immersed in a local world, not a world with modern ideas of freedom, not even the larger world of Rome, and in this small local world he was unwilling to be made to seem a fool. It is possible that his reluctance to publish arose from no deeper level than this.

By this time Copernicus was in his middle sixties. There was no medical service in those days capable of setting to rights all the small defects of the body which necessarily accumulate in all of us at such a time of life. The possibility is that by 1538 Copernicus was just plain tired and that the final revision of his manuscript represented a formidable and daunting task. Nevertheless, he agreed at last to undertake it. It is not clear whether this decision was taken before or after still another abstracted version, *Narratio prima de libris revolutionum,* was sent by von Lauchen in 1540 to his old teacher Johann Schöner in Nuremberg. The *Narratio prima,* written presumably with Copernicus' consent, appears to have had a favorable and somewhat sensational reception. Possibly this could have influenced Copernicus finally toward publication.

Von Lauchen extended his residence in Frauenburg into 1541, in the hope that eventually he would be able to return to Wittenberg with the finished manuscript. But it is usually said that he left Warmia with only the early mathematical chapters. The story now gains a peculiar momentum. Copernicus is said to have entrusted his manuscript to Tiedemann Giese, to make what arrangements he considered best. Giese is said to have sent it to von Lauchen. Then we are told that von Lauchen, having left Wittenberg to take up a post in Leipzig, entrusted Andreas Osiander to see the work through the press in Nuremberg. With the faithful von Lauchen thus removed, Osiander surreptitiously added a Preface of his own making, the Osiander Preface being in substance the same as the Osiander letter to Copernicus in 1541: ". . . hypotheses are not articles of faith. . . ."

As the late Walter Baade used to say, one can detect "a slight odor of rat" in this story. Are we to believe that von Lauchen, after making the long journey from Wittenberg to Frauenburg in the hope of learning something of the Copernican theory, after waiting there two years in the hope of acquiring the precious manuscript, would suddenly turn it over to Osiander? Is the journey from Leipzig to Nuremberg so great that von Lauchen could not have made that particular journey, even on a number of occasions? And is it not curious to find Copernicus writing to this same Osiander (1540) before there was any question of Osiander's involvement with the manuscript?

The curiosities do not end here. By a lucky stroke Copernicus' own manuscript, after being thought lost, was recovered in Prague in the second half of the nineteenth century. It became immediately clear that *De revolutionibus* had not been set from this manuscript. Quite a major editing job had been done between the manuscript and the printed version, editing which in retrospect must be considered improper. The manuscript is untitled, so we do not really know how Copernicus wished his book to be named. From additions to the text it has been speculated that the technical editing was done by a certain Erasmus Reinhold. But what has happened in the meantime to von Lauchen? The technical editing could have been done just as well in Leipzig as in Nuremberg.

In a first departure from this unsatisfactory story, I tried the idea that von Lauchen stayed for two years in Frauenburg to help Copernicus prepare a final manuscript. Suppose his help proved less than competent and that in the end Copernicus had to seek assistance elsewhere—from Osiander and Reinhold. This would explain many of the apparent contradictions. But an examination of the *Narratio prima* shows more brashness than incompetence. I also found the emendations and corrections in the Copernicus autograph to be firm and decisive, which seemed inconsistent with the idea of his having become too enfeebled to work alone.

At this stage I became impressed by the general similarity between the *Narratio prima* and *Commentariolus* (alternative geometrical constructions are not important here). If a brash young man had not hesitated to make what was essentially a copy of *Commentariolus,* would he not also have made a copy of the main work? On this basis, von Lauchen would have left Frauenburg after his two-year stay, not just with the early mathematical chapters of *De revolutionibus* but with his own personal copy of the whole book. And now, just as he had publicized his version of *Commentariolus,* so he arranged to print his version of *De revolutionibus.*

Copernicus finally released his autograph to Tiedemann Giese, and Giese very likely passed it to von Lauchen, just as the usual story has it. But by this time the book was already being printed from von Lauchen's amended copy. Under this interpretation we can easily understand the angry letter sent after the death of Copernicus by Giese to von Lauchen, complaining in strong language about the liberties taken with the book.

It is worth noting that von Lauchen's copy would quite likely not have contained a preface, since authors usually add the preface as a last touch. It is therefore possible that Osiander did *not* suppress the Copernicus Preface. In the absence of an author's preface he added an editorial preface in which he referred to Copernicus in the third person. Somewhat naturally, his preface was similar to his letter of 1541 to Copernicus. The sentiments it expresses agree remarkably well with the outlook of modern theoretical physics, and are not at all inept, as earlier generations have supposed.

The rest of the story is quickly told. The book appeared in March, 1543, but a copy did not reach Copernicus until May. He was then on his deathbed, and his actual death occurred within hours of the book's arrival.

To the citizens of Frauenburg it must only have seemed as if an infirm old man of some past distinction had passed from their midst. They were not to know that drums were already rolling with the noise of distant thunder. His fellow canons, as they laid him in his grave, were not to know that Copernicus had detonated an overwhelming explosion of human knowledge, an explosion still with us in our own day, and whose eventual outcome we cannot yet foresee.

III

THE HELIOCENTRIC THEORY
OF ARISTARCHUS

We saw in Chapter I that the simple heliocentric theory of Aristarchus leads to unacceptably large errors, particularly for Mars, even if we only compare the predictions of the theory with naked-eye observations. Yet to arrive at the picture of Aristarchus a considerable degree of sophistication is needed. In this chapter we shall see more closely what is involved.

There was little chance of ancient astronomers' arriving at a tolerable picture of the planetary motions until it had been realized that the Earth is essentially spherical in shape. Two observations which led to this conclusion are worth mentioning. When one journeys south, new stars not previously visible appear above the southern horizon, and stars dip lower toward the northern horizon. The Greeks knew long before Hipparchus that the constellation of the Plow comes down to the northern horizon when seen from Egypt but not when seen from Greece. Assuming the stars to be distant objects, this shows that the Earth is not flat. More directly, when the Earth comes between the Sun and the Moon, the shadow which the Earth throws onto the Moon is seen to be circular in shape. By judging the apparent radius of the shadow it is possible to determine the ratio of the distance to the Moon to the radius of the Earth, and Hipparchus succeeded in obtaining a surprisingly good estimate this way (the ratio is about 60). The spherical shape of the Earth was understood and accepted by the Greeks from the time of Plato, so there was no impediment to later astronomers on this score.

The diurnal rotation of the heavens is the most obvious motion to be fitted into our picture. It makes little difference from a purely kinematical point of view whether we consider the Earth to be fixed, with the heavens rotating, or vice versa, although dynamically it would be awkward (but not impossible) to think in terms of a rotation of the heavens. Indeed, kinematically it is somewhat simpler, at any rate to begin with, to think in terms of the heavens as rotating. We have the situation shown in Figure 4, where we think of the stars being projected onto a distant sphere with the center taken at the Earth, the celestial sphere. The celestial sphere is considered to rotate about an axis directed toward the pole star P. The circle in which the plane through the center of the Earth taken perpendicular to the polar axis PP₁ cuts the celestial sphere is called the celestial equator. The angle which a line drawn from our particular position on the surface of

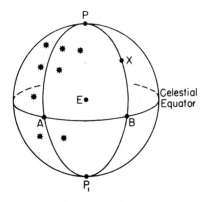

Figure 4. The determination of latitude and longitude on the celestial sphere. The latitude of the point X is the angle which the circular arc XB subtends at the Earth E; the longitude is the angle which AB subtends at E, A being a fixed reference point (see Figure 6). The celestial sphere rotates once in a day, the reference point A partaking in the motion. It follows that this diurnal rotation does not change latitudes and longitudes.

the Earth to the center makes with this plane is called our latitude.

We can obviously set up a system of latitude and longitude on the celestial sphere as well as on the surface of the Earth. This is what astronomers do in order to describe the positions of stars and other objects.

Longitudes are reckoned with respect to a standard position, just as longitudes on the Earth are reckoned with respect to a standard place (Greenwich). Notice that the rotation of the heavens does not affect latitudes and longitudes on the celestial sphere because the standard position is taken to partake in the diurnal motion.

For the Sun, Moon, and planets we determine latitudes and longitudes in the same way, but in these cases the latitude and longitude values change with time. If, at any particular moment, a planet, or the Sun or Moon, happens to be very close to a particular star, we assign essentially the same latitude and longitude to it that we do to the star, but because the planet or the Moon or Sun, as the case may be, is found to move with respect to the stars we have to be constantly changing the assigned latitudes and longitudes. The determination of how the latitudes and longitudes change with time is precisely the problem before us.

Of all such cases it is easiest to observe the motion of the Moon with respect to the stars. This shows up after a few nights of casual observation. If followed for a month or two, the Moon is found to move along a path on the celestial sphere of the kind shown in Figure 5. Careful observation over many years shows, however, that the Moon does not quite retrace its path from one month to the next. Starting with a particular path in a particular month, it takes 18.61 years before the

Moon returns to that same path, and in this time the angle which the plane of the path makes with the plane of the celestial equator goes through all values from about 19° to 29°, and back again to the starting value. This complex behavior was probably already known to Stone Age man. It plays a critical role in the occurrence

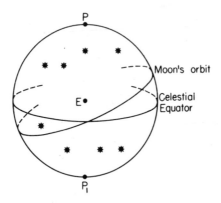

Figure 5. In addition to sharing the diurnal rotation of the celestial sphere, the Moon follows a path among the stars which it completes in a lunar month of 27.3 days.

of eclipses and for this reason has been important to astronomers throughout the ages. It must have seemed just as important to explain this curious behavior of the Moon as it seemed to explain the motions of the planets. Yet whereas the planetary problem lay reasonably within the grasp of early astronomers, the theory of the motion of the Moon was not accessible to them, or to Copernicus or to Kepler. The broad features of the lunar motion had to await Newton's theory, while the fine details were not subject to analysis until the nineteenth century—indeed, there are problems still extant.

The early astronomers could not know which problems they could hope to solve. Perforce they had to take a shot at everything. And because insoluble problems were mixed up with soluble ones their task in dealing with the soluble ones was made all the harder. This must always be remembered in attempting to understand the difficulties which beset Ptolemy and Copernicus. Both expended much effort in attempting to understand the Moon, whereas they would probably have gone further with less effort if they had ignored this problem.

The Sun is harder to observe than the Moon, because the glare of the Sun prevents the background of stars from being seen in full sunlight. Nevertheless, by observing stars close to the Sun immediately before sunrise and immediately after sunset it is possible to make a determination of the path of the Sun on the celestial sphere. It is found that the latitude and longitude of the Sun also change from day to day, and that the Sun returns to its original position in a year. The path followed by the Sun does not change significantly from year to year, so fortunately we do not have to face the complexities which bedevil the motion of the Moon. The path of the Sun projected on the celestial sphere is shown in Figure 6. The plane of the path makes an angle of about 23.5° with the plane of the celestial equator. It is this tilt which explains the seasons of the year. For the Northern Hemisphere of the Earth there is midwinter when the Sun is at B, spring at γ, summer

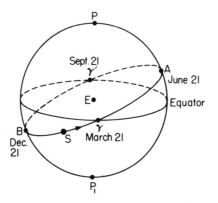

Figure 6. The Sun follows an annual path among the stars, the plane of the path making an angle of about 23.5° with the plane of the equator. The Sun's orbit cuts the equator at γ, γ'. The point γ is taken as the reference point for measuring longitudes on the celestial sphere. That is to say, the point γ in this figure is the point A of Figure 4.

at A, and autumn at γ'. The points γ and γ' are the nodes of the Sun's orbit, γ being called the first point of Aries. Longitudes of stars on the celestial sphere, as well as longitudes of planets and of the Sun and Moon, are reckoned with respect to γ.

It makes no difference, kinematically, whether we think of the Sun's observed path among the stars as

being due to a motion of the Earth around the Sun or
of the Sun around the Earth. For the moment, then, let
us follow the old belief that the Sun moves around the
Earth. The path drawn in Figure 6 is a circle because
it is a projection of the orbit onto the celestial sphere.
The actual orbit in space need not be a circle, and

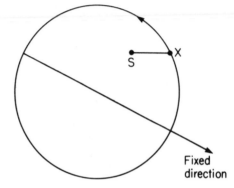

Figure 7. Hipparchus' theory of the seasons. As X moves
around the circle, the segment drawn from X to the Sun S
makes a constant angle with a fixed direction.

indeed must not be a circle because, if it were, summer
and winter would be equal, whereas the seasons are
found to be unequal. It was already known to the
Greeks that spring-to-summer-to-autumn differs from
autumn-to-winter-to-spring by three days. Indeed, this
fact was probably known long before the Greeks. It
was explained by Hipparchus in the manner of Figure
7, where the point X moves uniformly in a circle

around the Earth in a year. The Sun S is fastened to X by a short stick (not necessarily a real stick) which maintains a constant direction throughout the annual motion of X. Ugly and implausible or not, the Hipparchus theory grapples with the facts whereas the circular picture of Aristarchus fails to do so. It will

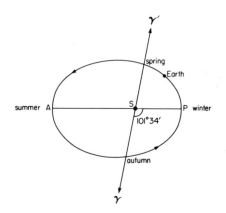

Figure 8. P is perihelion point, A is aphelion, and γ is the first point of Aries.

be clear then why it was so hard to arrive at the basic elements of a heliocentric theory. The two most striking bodies in the sky, the Sun and Moon, cause difficulties at the outset, even before we come to the planets. To make progress we have to be courageous enough (or foolhardy enough) to ignore the complexities of the Moon's orbit and also the known inequality of the seasons. As an aside, before we proceed, it may be of interest to compare the Hipparchus theory of Figure 7 with Figure 8, which gives the modern view of the inequality of the seasons.

If the reader now refers back to Plate I, he may well wonder how a knowledge of the planetary motions can possibly make things any easier. Actually it does, as we shall see in the remainder of this chapter. First, there is the welcome simplification that the orbits of the planets are all more or less in the same plane as the path of the Sun. The measured tilts of the planes of the planetary orbits to the plane of the Sun's orbit for the six planets known to Copernicus and to the Greeks are given, together with other quantities of interest, in Table 3.*

TABLE 3

Planet	Inclination to Sun's orbit	Period of revolution (years)	Longitude of perihelion	Distance (relative radius)
Mercury	7° 0′	0.2408	76° 13′	0.387
Venus	3° 24′	0.6152	130° 27′	0.723
Earth	—	1.0000	101° 34′	1.000 (standard)
Mars	1° 51′	1.8808	334° 35′	1.524
Jupiter	1° 18′	11.862	13° 2′	5.203
Saturn	2° 29′	29.457	91° 29′	9.539

The first step in breaking into the problem is to consider the observations of Venus. Venus sometimes

* The gravitational influences of one planet on another cause their orbits to change slowly. The values in Table 3 apply at January 1, 1920.

runs ahead of the Sun in its path and sometimes it lags
behind, in a kind of oscillatory motion, which we might
try to represent in the manner of Figure 9, with Venus
swinging in harmonic motion along a line through the
Sun. In the part of the cycle from R to L, Venus would
move more rapidly forward than the Sun, from L to R
it would move less rapidly. At R the planet would be

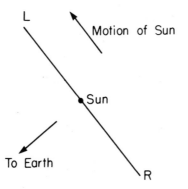

Figure 9. The apparent motion of Venus, as seen from the
Earth, consists of an oscillation backward from L to R, then
forward from R to L.

behind the Sun, at L it would be ahead. This picture
therefore corresponds qualitatively with the observational situation. It fails, however, when we take account
of quantitative aspects of the data. On the model of
Figure 9 we would expect variations in the brightness
of Venus in the half-cycle from L to R to be exactly
matched by similar variations in the opposite half-cycle
from R to L. This is incorrect. The two parts, L to R

and R to L, are not the same, suggesting that the distance from the Earth to Venus is not the same. Nor, indeed, is the time of swing from L to R equal to the time from R to L, as it would be for simple harmonic motion.

A more sophisticated picture of the motion of Venus

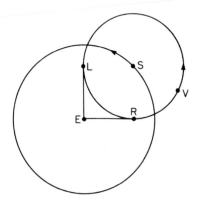

Figure 10. This model of the motion of Venus provided a first breakthrough in the understanding of the planetary motions. When projected onto the celestial sphere it gives the apparent oscillation of Figure 9.

is shown in Figure 10. We now have a situation in which the Sun S moves uniformly in a circle around the Earth (ignoring the inequality of the seasons) and in which Venus V moves uniformly around the Sun in another circle. At point L of this second circle Venus is moving directly toward the Earth, at R Venus is moving away from the Earth. The situation projected on the sky is like Figure 9, but now the distance of Venus from the Earth is different over the sections L to R and R to L, and the times are also different. Also, we have avoided the embarrassment that Figure 9 would require Venus to pass through the Sun.

The relative-size relation of the two circles of Figure 10 is readily determined by measuring the angle between the direction of the Sun and the direction of Venus at the moment when Venus is most ahead of the Sun. A similar angle when Venus is most behind the Sun can also be measured, and the sum of these two angles is equal to the angle LER of Figure 10. With this known, the ratio of the radius SV of the moving circle to the radius ES can be determined. It turns out to be about 0.723, the entry for Venus in the relative radius column of Table 3. The time for the motion L to R added to the time back again from R to L gives the synodic period of Venus in its orbit, 1.599 years.

We now have a model with predictive capacity. Although designed to fit a set of known data, it can be used to explain further new sets of data. This is a crucial requirement of a useful theory. Theories which only explain already known facts usually end up in a blind alley. The predictions of the model turn out to be quite good, for a reason we can understand from Table 1 of Chapter I. The eccentricity of the orbit of Venus happens to be very small, and the eccentricity of the Earth's orbit is reasonably small. So the actual orbits are nearly circles, which means that Figure 10 is a good approximation to the real situation. This regularity was the first to be discovered among the planets. It was known to Heraclides and was therefore available to Hipparchus and Ptolemy as well as to Aristarchus. From a code-cracking point of view, the case of Venus is a breakthrough.

The situation for Mercury is similar to that for Venus, but because of the large eccentricity of the orbit of Mercury the approximation given by a simple model like Figure 10 is not nearly so good. Fortunately, the shortcomings of a simple model were not so obvious in really ancient times, because reasonably accurate ob- servations were not easy to make on account of the closeness of Mercury to the Sun. However, the dis- crepancies were well known to Ptolemy, who had con- siderable difficulty in improving on Figure 10.

The outer planets behave in a way that is both curiously similar to and curiously different from the case of Venus. Imagine the Sun removed from Figure 10. We should not then be able to compare observation- ally the direction E to V with the direction of the Sun in the same way as before. We should be aware, how- ever, that the arcs L to R and R to L were very different, because over the arc L to R the planet would appear to go backward in the sky, while over R to L it would appear to go forward. This alternation of backward and forward motions, with the forward motions winning on balance, is just what is observed for the outer planets. We look therefore to a model of the kind drawn in Figure 11, but with C now only a geometrical point, since the outer planets are not observed to oscillate about the position of the Sun. As in the case of Venus, we seek to find the angle LER by comparing the lengths

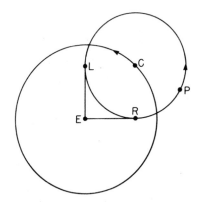

Figure 11. This model applies to the motions of planets more distant from the Sun than the Earth. It is like Figure 10 for Venus, except that the Sun is no longer at the center of the small circle.

of the backward and forward motions. The matter is not so straightforward as before, however, because C is not a visible point. Nevertheless, from a thorough analysis of the data the angle LER can be determined. This again fixes the ratio of the radius CP of the small circle to the radius EC of the large circle. And the periods of time in which C moves around the large

circle and P about the small circle can be obtained from
a suitably complete set of data. Once again we arrive
at a model with predictive capacity. The results turn
out to be reasonably good for Jupiter and Saturn, but
much less good for Mars—partly because the eccentricity of the actual orbit of Mars is about twice as great
as the eccentricities for Jupiter and Saturn, and partly
because the Earth and Mars approach each other much
more closely.

So far, the route toward understanding the motions
of the planets is clearly marked—with the solar system
being the way it is, there is no other way of making
a sensible start on the problem. But now we have
alternative routes ahead. Which we elect to follow will
depend on our temperament. If we are overwhelmingly
concerned with the predictive quality of our model
(which would be the point of view of most modern
theoretical physicists), we shall seek to modify Figure
11 in order to secure a better correspondence between
theory and observation. This is just what Ptolemy
sought to do. But if we are concerned not so much with
quantitative accuracy as with the qualitative structure
of the problem, we shall be unable to overlook two
further strange aspects of the situation.

It turns out that P in Figure 11 is required to move
around the small circle in a year, the same period as S
moves around E in Figure 10. But why should the
Sun be involved with Figure 11 at all? Here we are

concerned with the relation of the planet P, whether Mars, Jupiter, or Saturn, to the Earth. Indeed, in Figure 11 we were required to *remove* the Sun from the center of the small circle. Still more curious perhaps, if we ask where the Sun should be placed with respect to Figure 11, it turns out as in Figure 12: from observa-

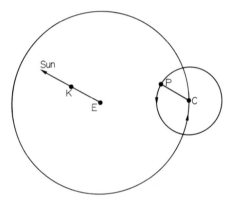

Figure 12. The Sun is observed to lie along a line through E drawn parallel to CP. The point K is chosen so that EK=CP.

tion it is found that a line drawn from the Earth in the direction of the Sun is always parallel to the line joining C to P.

Clearly this must mean *something*, but we cannot decide *what* from observation alone. Indeed, if observation had resolved this issue, there would have been no controversy in astronomical history between geocentric and heliocentric theories. To distill meaning out of these

curious facts we must take a theoretical step. Some-
where along the line EK of Figure 12 lies the Sun, but
observation does not tell us where (at any rate observations available to the Greeks or to Copernicus). The
inspired step taken by Aristarchus, and later inde-
pendently by Copernicus, was to place the Sun at K,

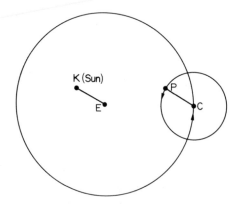

Figure 13. This is the same diagram as Figure 12, but with
the Sun placed by hypothesis at the point K. This was the
crucial step needed to arrive at a heliocentric theory.

where K is chosen such that the length E to K is equal
to the length C to P. It is worth redrawing Figure 12
with the Sun S placed firmly in this position, as in Figure
13, in order to illustrate the geometrical fact that the
motion of P described by Figure 13 is exactly the same
as that described in Figure 14, where the planet P now
goes in a larger circle about S.

It may take the reader a moment or two to become

satisfied that Figures 13 and 14 do indeed give the same motion of P. The equivalence of these two pictures was already known to Apollonius, who lived in the third century B.C., long before Ptolemy (*ca.* A.D. 150). The reader familiar with complex numbers will note

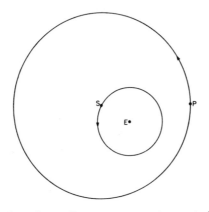

Figure 14. The planet P now moves in a circle about S, while S moves in a circle about E. The motion of P is exactly the same as in Figure 12.

that if E is taken as origin and Z_1, Z_2 are complex numbers determining the points S, C respectively, then P in Figure 14 is $Z_1 + Z_2$, while P in Figure 13 is $Z_2 + Z_1$. Because complex numbers are commutative the positions of P are the same in the two cases. In the next chapter we shall find that the constructions discovered by Ptolemy and Copernicus can be understood in an elegant way by using complex numbers.

We have a construction of the form of Figure 10 for

an inner planet and one of the form Figure 14 for an outer planet. If they are drawn together, as in Figure 15, we can at last take advantage of the fact that the motion of S around E is kinematically equivalent to the motion of E around S. Thus Figure 15 is equivalent

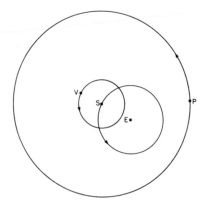

Figure 15. This combines Figures 10 and 14, with V representing an inner planet and P representing an outer planet.

to Figure 16. The respective circles have the same radii in the two figures, so that with the use of the same periods of motion in corresponding circles the relative positions of E, S, V, P must be the same in the two pictures at all times. Here P can represent any one of the outer planets and V can represent either Venus or Mercury. For the six planets observed in ancient times, using the relative radii of Table 3, we get the structure shown in Figure 17. This is the heliocentric theory of Aristarchus.

Copernicus probably rediscovered the line of reasoning set out above before returning from Italy to Poland in 1503. I suggested in the previous chapter that the year 1513 may well have represented a watershed both

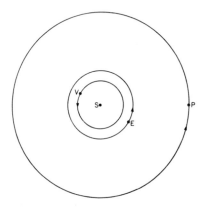

Figure 16. The motions in this heliocentric picture are exactly the same as in Figure 15, because the motion of S about E in Figure 15 is kinematically equivalent to the motion of E about S in this figure.

in his life and in the development of the heliocentric theory. The tendency to suppose that Copernicus discovered his theory early in life arises, I believe, from a misjudgment of the magnitude of his achievement. The popular view, and I think to some extent the view of the less technical biographies, is that the simple

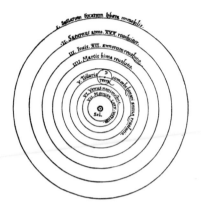

Figure 17. The heliocentric picture for the six planets known in the time of Copernicus, as it was drawn in *De revolutionibus.*

heliocentric theory described above represents the major part of what he did. So far from this being the case, we have as yet covered only the easy beginning. For Copernicus, the hard part, the fight with Ptolemy, had still to come.

I V

COPERNICUS AND PTOLEMY

$\mathbb{B}$OTH Copernicus and Ptolemy arrived at their constructions from the analysis of a vast amount of data. Simply to describe the constructions here without including explicit comparison with observation would be a somewhat pointless procedure. We should be involved in geometrical intricacies whose *raison d'être* was unclear. The need to include long and extensive reference to the data can be avoided, however, if we start from the elliptic orbit given in equation (6) of Chapter I, for we know that such an orbit represents all the data correctly (at any rate to within the margins we are interested in here). So we can convert the question of how the constructions of Ptolemy and Copernicus compare with observation to the question of how the constructions compare with the elliptic orbit. This is the approach to be followed in the present chapter. Instead of following historic evolution, in which data are used in a struggle toward a more correct theory, we start with the correct theory and see to what extent it has to be degraded in order to arrive at the methods used by Ptolemy and Copernicus.

The starting point then is the equation

$$r = \frac{a(1 - e^2)}{1 + e \cos \theta} \tag{7}$$

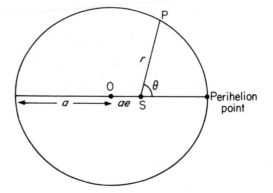

Figure 18. The angle θ is measured from the perihelion point. The distance from the center O to the focus S is ae, and the distance from S to P is r.

in which r is distance measured from the Sun, e is the eccentricity of the ellipse, and θ represents the angle between the major axis and the line drawn from the Sun to the planet. The relation of the focus to the center of the ellipse is shown in Figure 18, the distance being the product ae. The distance of the planet from the Sun at perihelion is $a(1 - e)$ and at aphelion $a(1 + e)$.

We also need Kepler's law of equal areas. Suppose in

a small time interval dt the line from S to P turns through an element of angle $d\theta$. If r were to remain unchanged, the line from the Sun to the planet would sweep out a narrow isosceles triangle of height r and of small base $r\,d\theta$, and so would have area $\frac{1}{2}r^2\,d\theta$. Even if r changes a little in the time dt, the area remains $\frac{1}{2}r^2\,d\theta$ to within the first order of smallness. The rate at which the planet sweeps out an area is therefore $\frac{1}{2}r^2\dfrac{d\theta}{dt}$; any small term due to changing r becoming negligibly small in the limit as dt tends to zero. It follows from Kepler's law that $r^2\,d\theta/dt$ is a constant,

$$r^2\frac{d\theta}{dt} = h, \tag{8}$$

where h is a constant. The mean motion n was defined in equation (3) of Chapter I. According to this equation the time required for a planet to make a complete revolution in its orbit is $2\pi/n$. In this length of time the total area swept out by the line from S to P is

$$\frac{1}{2}r^2\,\frac{d\theta}{dt}\cdot\frac{2\pi}{n} = \pi\,\frac{h}{n}\ , \tag{9}$$

which must be equal to the area πab of the whole ellipse. Using equation (1) of Chapter I, we can write πab in the form $\pi a^2(1 - e^2)^{1/2}$, and this is equal to equation (9),

$$\pi a^2(1 - e^2)^{1/2} = \pi\,\frac{h}{n}, \tag{10}$$

or simply

$$n = \frac{h}{a^2(1 - e^2)^{1/2}} \qquad (11)$$

for the relation of the mean motion to h, a, e.

Eliminating r between equations (7) and (8) gives

$$\frac{d\theta}{dt} = \frac{h}{a^2(1 - e^2)^2}[1 + e \cos \theta]^2$$

$$= \frac{n}{(1 - e^2)^{3/2}}[1 + e \cos \theta]^2 . \qquad (12)$$

If we could conveniently integrate this equation to determine θ in terms of the time t, obtaining some explicit function, $\theta(t)$ say, we should have a closed solution to the problem of determining the motion with respect to time. $\theta(t)$ would give the heliocentric longitude at time t, while the substitution of $\theta(t)$ in equation (7) would give the heliocentric distance. This cannot be done, however. Equation (12) can only be integrated as an infinite series, indeed the series given already in equation (2) of Chapter I:

$$\theta = nt + 2e \sin nt + \frac{5}{4} e^2 \sin 2nt + \dots \qquad (13)$$

Substitution of equation (13) in equation (7) gives the following infinite series for r:

$$\frac{r}{a} = 1 - e \cos nt + \frac{1}{2} e^2(1 - \cos 2nt) + \cdots \qquad (14)$$

We already remarked in Chapter I that it is this feature of the correct solution which made the problem of determining the planetary motions so awkward.

Suppose we represent the planet by a complex number:

$$z = r \exp i\theta$$
$$= a(1 - e \cos nt + \ldots) \exp i(nt + 2e \sin nt + \ldots).$$
$$(15)$$

Now degrade the problem by working only to the first order in the eccentricity e. It is then sufficient to write

$$\exp (2ie \sin nt) \doteq 1 + 2ie \sin nt , \qquad (16)$$

and for equation (15) we have

$$z \doteq a(1 - e \cos nt + 2ie \sin nt) \exp int$$

$$= a(1 + \frac{1}{2} e \exp int - \frac{3}{2} e \exp - int) \exp int$$

$$= - \frac{3}{2}ae + a \exp int + \frac{1}{2}ae \exp 2int. \qquad (17)$$

Evidently the position P of the planet can be represented in this order of approximation by the following rules:

1. Make a displacement of length $\frac{3}{2}ae$ from S to a point K in a negative sense along the real axis.

2. Then make a displacement $a \exp int$ from K to a point L.

3. Finally, make a displacement $\frac{1}{2}ae \exp 2int$ from L to P.

These rules are shown geometrically in Figure 19. The point K stays fixed. As t changes, the point L moves with angular velocity n about K in a circle of radius a, the sense of the motion being anticlockwise, while the planet P itself moves about L with angular velocity $2n$ in a circle of radius $\frac{1}{2}ae$, also in an anticlockwise sense. The angular velocities here are measured, of course, with

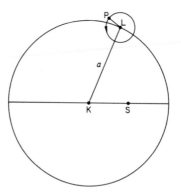

Figure 19. This is the construction of Copernicus. The line KL turns at an angular rate equal to the mean motion n of the planet, whereas LP turns at an angular rate $2n$. The length LP is 1/3 of KS, which is $3ae/2$.

respect to a fixed direction—the real axis of our complex numbers. This is the construction of Copernicus.

Actually, Copernicus made two minor deviations from these rules, deviations which were not improvements, and which were subsequently corrected by Kepler in his earliest attempts on the problem. The small circle from L to P was omitted in the case of the Earth. Copernicus also replaced the Sun S, in all planetary constructions other than that of the Earth, by the point K determined for the Earth. This meant that the Copernican theory of the planetary motions was not strictly helio-centric to the first order in the eccentricity of the Earth's orbit.

It has to be remembered, however, that Copernicus felt called on to face up to every problem contained in the data available to him. In the previous chapter we saw how he became concerned with the difficult problem of the motion of the Moon. It is worth interrupting the present discussion for a moment to give a second example.

The planetary orbits are not strictly ellipses, as we have so far taken them to be, because one planet disturbs the orbit of another through the gravitational force which it exerts. For the most part these effects are slight, but the fact that the line S to K of Figure 19 slowly changes for the Earth's orbit was known to Copernicus, from a comparison of the observations of Arabic astronomers with those of the Greeks. As always, Copernicus felt obliged to include even this minor effect in his kinematical description. From certain other comparisons it also seemed to him as if the Earth's whole orbit were subject to oscillation, "trepidations" as he called them. These effects, later shown by Tycho Brahe to be illusory, did not survive to trouble Kepler. But they were a serious worry to Copernicus, who found himself trying to represent, in as simple a way as possible, a considerable catalogue of small motions, some of them merely artifacts of bad data. In retrospect, it is of course easy for us to see that Copernicus should have been less punctilious. He would have done better simply to have applied the construction of Figure 19 consistently to all planets—but this is to be wise after the event.

Returning now to our main argument, equation (17) can be rewritten in the form

$$z \doteq a(1 - e \cos nt + 2ie \sin nt) \exp int$$
$$= a(1 + e \cos nt) \exp int$$
$$\quad -2ae(\cos nt - i \sin nt) \exp int$$
$$= a(1 + e \cos nt) \exp int - 2ae. \qquad (18)$$

Hence we can also arrive at the position of a planet, again correct to the first order in the eccentricity, by using the following alternative rules:

1. Make a displacement of length $2ae$ from S to a point A in the negative sense along the real axis.
2. Make a displacement of length $a(1 + e \cos nt)$ from A in a direction making an angle nt with the positive real axis.

How shall we interpret this construction with respect to changing time? Turning to Figure 20 we see that the position P for the planet lies, to the first order in the eccentricity, on the circle of radius a with center C at the mid-point of AS. This is because the distance from A to P is $a(1 + e \cos nt)$, the angle PAS being nt. This alternative way of representing the motion of the planet uses only one circle. The odd feature is that the radius

C to P does not turn at a uniform rate. It is the line A to P which turns uniformly. Copernicus was aware that the motion could be represented this way, but he disliked the concept of turning about the center of a circle at a nonuniform rate. Kepler had no such scruples. He seems to have preferred Figure 20 to Figure 19.

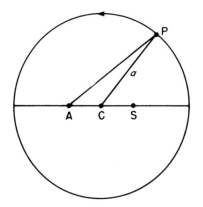

Figure 20. This is the basic construction of Ptolemy, but applied in a heliocentric picture. Here AC = CS = ae (these distances are exaggerated), and the line from A to P turns at an angular rate equal to the mean motion n.

Except that it is related to motion about the Sun instead of about the Earth, Figure 20 is the construction of Ptolemy. The point A is his *punctum aequans*. Write the position of the Earth relative to the Sun in the form

$$z_E \doteq a_E[1 + e_E \cos n_E(t - t_E)] \exp i n_E(t - t_E) - 2a_E e_E,$$

$$(19)$$

and write a similar equation for a planet P:

$$z_P \doteq a_P[1 + e_P \cos n_P t] \exp i n_P t - 2a_P e_P. \qquad (20)$$

Notice that from the beginning, in equation (13), we have taken $\theta = 0$ at $t = 0$. This meant that throughout the above discussion of the motion of a single planet we measured the time t from a moment when the planet was at perihelion. Now that we are considering two planets, E and P, we cannot suppose that both were simultaneously at perihelion. According to equation (19), E is at perihelion at $t = t_E$, and according to equation (20), P is at perihelion at $t = 0$.

The position of P relative to E is given by the simple subtraction of equation (19) from equation (20):

$$z_P - z_E \doteq a_P[1 + e_P \cos n_P t] \exp i n_P t - 2a_P e_P$$
$$- a_E[1 + e_E \cos n_E(t - t_E)] \exp i n_E(t - t_E) + 2a_E e_E .$$
$$(21)$$

The form of equation (21) is not as straightforward to handle geometrically as equation (18), and to overcome the added complexity occasioned by this subtraction Ptolemy was forced to degrade his theory slightly by omitting further small terms. He would have done better by working with equation (18), i.e., by using a heliocentric theory. Reference back to Table 1 of Chapter I shows that the eccentricities of all planetary orbits but that of Venus are large compared to the eccentricity

e_E of the Earth's orbit. Suppose for all planets except Venus we omit terms involving e_E from equation (21):

$$z_P - z_E \doteq a_P\{(1 + e_P \cos n_P t) \exp in_P t - 2e_P\}$$
$$- a_E \exp in_E(t - t_E). \quad (22)$$

We can represent the first term here by exactly the same construction as that of Figure 20. So to obtain P relative to E, according to equation (22), we only have to add to Figure 20 the effect of the second term. Noting that

$$- a_E \exp in_E(t - t_E) = a_E \exp in_E(t + \pi - t_E), \quad (23)$$

the required construction has the form given in Figure 21. We now have two circles, as in the Copernicus construction of Figure 19. The point L moves about the circle of center C and radius a_P, but with constant angular velocity n_P about the *punctum aequans* A, and the planet P moves around a circle of radius a_E and center L with angular velocity n_E. The position of P on this circle has phase $\pi - t_E$ at $t = 0$.

To use the Ptolemaic construction of Figure 21 one needs in practice to determine the following quantities for each planet:

the ratio a_P/a_E,
the eccentricity e_P,

a time when the planet is at perihelion. This is the same information as is needed to make practical application of the Copernicus construction of Figure 19. The a_P/a_E

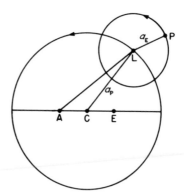

Figure 21. The construction of Ptolemy applied to a geo-centric picture. The line AL turns with the mean motion n_P of the planet, while the line LP turns with the mean motion n_E of the Earth.

values used by Ptolemy and Copernicus are given in Table 4.

TABLE 4

Planet	a_P/a_E Ptolemy	a_P/a_E Copernicus	a_P/a_E Modern
Mercury	0.3708	0.3763	0.3871
Venus	0.7194	0.7193	0.7233
Earth	—	1.0000	1.0000
Mars	1.5191	1.5198	1.5237
Jupiter	5.2164	5.2192	5.2028
Saturn	9.2336	9.1743	9.5388

There is no entry in the first column for the case of the Earth because the motion of the Sun in the Ptolemaic theory is represented by $-Z_E$, and Z_E, given by equation

(19), is represented by the construction of Figure 20, in which there is only one circle.

In all cases in Table 4 the difference between the modern value and the value used by Copernicus is in the same sense as the value used by Ptolemy, which indicates the extent to which Copernicus relied on Ptolemy's data. Copernicus evidently used Ptolemy's values as a working basis, modifying them somewhat to accord with his own observations and with those of Arabic astronomers. The situation seems to have been the same for the e_P values used in the two theories.

From what has so far been said it will be realized that the predictive capacity of the constructions of Ptolemy and Copernicus are very nearly the same. Indeed, the error in the Copernican theory occasioned by taking the point K determined for the Earth's orbit as the point S for the other planets meant that the Copernican theory involved just the same approximation as we made in passing from equation (21) to equation (22). Terms of order $a_E e_E$ were neglected in both theories. Copernican theory becomes superior to Ptolemaic theory, however, when account is taken of the inclinations of the planetary orbits (see Table 3, Chapter III). Copernicus took all the planes of the planetary orbits to pass through the point K for the Earth's orbit. Since K lies close to the Sun this was nearly the same as taking the Sun to lie in the plane of every orbit, which is the correct situation. Ptolemy, on the other hand, took all the planes to pass through the Earth, which is a worse approximation. Consequently, the Copernican theory of the motions of the planets in latitude is better than the Ptolemaic theory.

I have not sought in the above discussion to deviate from standard accounts of the two theories (for example, J. L. E. Dreyer, *A History of Astronomy from Thales to Kepler,* Dover Publications, 1953). I have simply interpreted these discussions in a modern sense through working directly from the elliptic orbits of the planets. At this stage I wish to take issue, however, with a criticism which is sometimes made of Ptolemy. It is said that Ptolemy's construction for Venus is a weakness of the theory, since it differs from Figure 21. But the change of construction for Venus shows that Ptolemy knew better than his critics what he was doing. It will be recalled that we passed from equation (21) to equation (22) on the basis that e_E, the eccentricity of the Earth's orbit, is smaller than e_P, the eccentricity of the planet's orbit. We remarked that this is true for every planet except Venus. To obtain a corresponding construction for Venus we must therefore work by omitting the e_P terms in equation (21), writing

$$z_P - z_E \doteq - a_E\{[1 + e_E \cos n_E(t - t_E)] \exp i n_E(t - t_E) \\ - 2e_E\} + a_P \exp i n_P t \quad (24)$$

in place of equation (22). The construction for equation (24) is like Figure 21, but with EC = AC = $a_E e_E$ instead of $a_P e_P$, with a_E for CL instead of a_P, and with a_P for LP instead of a_E. The construction has become inverted with respect to E and P.

This ends our interpretation of the work of Ptolemy and Copernicus. It seems truly remarkable in retrospect that Ptolemy did not realize that the second term in equation (22), which is the term giving rise to the epicyclic motion of P about L in Figure 21, could be interpreted as the Earth's motion reflected in the motions of the planets. This is clearly seen in the above treatment, for this particular term appeared when we subtracted the position of the Earth Z_E from the position of the panet Z_P. It would be valuable, as a matter of historical interest, to go over Ptolemy's work (in as near to its original form as possible) to find if there are any hints as to what his position really was. Did he really overlook the source of the epicycle in Figure 21? Or was he so overwhelmingly concerned with the predictive capacity of his theory that the issue did not concern him very greatly? I am led to wonder how far editors of the *Almagest*, like the editor of *De revolutionibus*, may have inserted their own interpretations and prejudices into Ptolemy's work at just this point.

EPILOGUE
FROM THE TWENTIETH CENTURY

$\mathbb{A}$т the beginning of Chapter I it was stated that we can take either the Earth or the Sun, or any other point for that matter, as the center of the solar system. This is certainly so for the purely kinematical problem of describing the planetary motions. It is also possible to take any point as the center even in dynamics, although a recognition of this freedom of choice had to await the present century. Scientists of the nineteenth century felt the heliocentric theory to be established when they determined the first stellar parallaxes. The positions of nearby stars were found to undergo annual oscillations, which were taken as reflections of the Earth's annual motion around the Sun. But, kinematically speaking, we can always give to the stars epicyclic motions similar to the ones we found for the planets in Chapter IV. Indeed, if we wish to consider the Earth to be at rest, it will be necessary to give an annual epicyclic motion to every object in the distant universe, as well as to the planets of the solar system. We cannot dismiss such a procedure simply on the grounds of inconvenience or absurdity. If our feeling that the Earth really goes around the Sun, not the Sun around the Earth, has any objective validity, there must be some important physical property, expressible in precise mathematical terms, which emerges in the heliocentric picture but not in a geocentric one. What can this property be?

Consider the well-known Newtonian equation

$$\text{mass} \times \text{acceleration} = \text{force}.$$

The mass for any particular body is intended to be always the same, independent of where the body is situated or of how it is moving. Suppose we describe the position of a body as a function of time in some given reference frame, and suppose we know the mass. Then, provided we also have explicit knowledge of the force acting on the body, Newton's equation gives us its acceleration. Determining the motion from there on is simply a mathematical problem—in technical terms we have to integrate the above equation. This procedure, which forms the basis of Newtonian mechanics, fails unless we know the force explicitly. In the Newtonian theory of the planetary motions, the theory leading to the basic ellipse from which we worked in Chapter IV, the force is taken to be given by the well-known inverse law:

Two masses, m_1 and m_2, distant r apart, attract each other with a force

$$G \frac{m_1 m_2}{r^2}$$

where G is a numerical constant. The force is directed along the line joining the bodies.

Now comes the critical question:

In what frame of reference is this law considered to operate?

In the solar system we cannot consider the inverse-square law to operate *both* in the situation in which the Sun is taken as the center and in that in which the Earth is taken as the center, because Newton's equation would then lead to contradictory results. We should find a planet following a different orbit according to which center we chose, and a body cannot follow two paths (at any rate not in classical physics). It follows that in order to use the inverse-square law in a constructive way we must make a definite choice of center. The situation which now emerges is that to obtain results that agree with observation we must choose the Sun as the center. If the Earth were chosen instead, some law of force other than the inverse-square law would be needed to give motions that agreed with observation.

Although in the nineteenth century this argument was believed to be a satisfactory justification of the heliocentric theory, one found causes for disquiet if one looked into it a little more carefully. When we seek to improve on the accuracy of calculation by including mutual gravitational interactions between planets, we find—again in order to calculate correctly—that the center of the solar system must be placed at an abstract point known as the "center of mass," which is displaced quite appreciably from the center of the Sun. And if we imagine a star to pass moderately close to the solar system, in order to calculate the perturbing effect correctly, again using the inverse-square rule, it would be essential to use a "center of mass" which included the star. The "center" in this case could lie even farther away from the center of the Sun. It appears, then, that the "center" to be used for any set of bodies depends on the way in which the local system is considered to be isolated from the universe as a whole. If a new body is added to the set from outside, or if a body is taken away, the "center" changes.

A similar circumstance was already present throughout our calculations, when we regarded angles as being measured with respect to a "fixed direction," it being implied that distant stars had directions that were "fixed" in this sense. If we make a calculation, using both Newton's equation and the inverse-square law, but measuring angles with respect to a direction that rotates with respect to the distant universe, things go very wrong. Newton was fully aware that his system of dynamics would work correctly only provided the "fixed directions" in the theory were chosen in a suitable way. His reference to the well-known rotating-bucket experiment was intended to illustrate this point.

It is clear therefore that in order to define the appropriate "center" of the local system in a useful way, and in order to define "fixed directions" relative to which angles are to be measured, we must take account of the relation of the local system to the universe outside. It seems that the local laws of force take simple forms only when the center is unaccelerated with respect to a frame of reference determined by the universe in the large, and when the fixed directions do not rotate with respect to the distant universe. From this point of view we can compare the heliocentric and geocentric theories of the solar system in an unequivocal way. We ask:

> Is it the Sun that is accelerated with respect to the universe, or is it the Earth?

Neglecting small effects, the answer is that the Earth is accelerated, not the Sun. Hence we must use the heliocentric theory if we wish to take advantage of simple rules for the local forces. But this is not to say that we cannot use the geocentric theory if we are willing to use more complex rules for the forces.

The present discussion has been formulated from the standpoint of the Newtonian theory, which is not well suited to problems concerning the universe in the large. We might hope therefore that the Einstein theory, which is well suited to such problems, would throw more light on the matter. But instead of adding further support to the heliocentric picture of the planetary motions, the Einstein theory goes in the opposite direction, giving increased respectability to the geocentric picture. The relation of the two pictures is reduced to a mere coordinate transformation, and it is the main tenet of the Einstein theory that any two ways of looking at the world which are related to each other by a coordinate transformation are entirely equivalent from a physical point of view. Moreover, in the Einstein theory the method of calculating the effect of gravitation is changed to a form which applies equally to all such related ways of expressing a problem.

It may still happen that it is easier to work through the details of a particular problem with respect to one coordinate system rather than to another, but no special physical merit is to be adduced from such a circumstance. Indeed, from a mathematical point of view, the problem of the planetary motions certainly continues to be easier to grapple with in the heliocentric picture. The simplication of such a picture shows itself in the

Einstein theory through boundary conditions which are impressed on the space-time structure at a large distance from the Sun—which is to say in terms of the control imposed by the universe in the large.

So we come back full circle to what was said at the beginning of this book. Today we cannot say that the Copernican theory is "right" and the Ptolemaic theory "wrong" in any meaningful physical sense. The two theories, when improved by adding terms involving the square and higher powers of the eccentricities of the planetary orbits, are physically equivalent to one another. What we can say, however, is that we would hardly have come to recognize that this is so if scientists over four centuries or more had not elected to follow the Copernican point of view. The Ptolemaic system would have proved sterile because progress would have proven too difficult.

BIBLIOGRAPHY

Armitage, A., *The World of Copernicus* (New York: New American Library, 1951).

Copernicus, N., *De revolutionibus orbium caelestium libri VI*, facsimile edition (Basingstoke: Macmillan, 1972).

Dreyer, J. L. E., *A History of Astronomy from Thales to Kepler* (London: Dover Publications, 1953).

Mizwa, Stephen P., *Nicholas Copernicus* (Port Washington, N.Y.: Kennikat Press, 1943, 1969).

Prowe, L., *Nicolaus Coppernicus* (Berlin, 1883–84).

Rosen, E., *Three Copernican Treaties* (London: Dover Publications, 1959).

Rybka, E., *Four Hundred Years of the Copernican Heritage* (Cracow, 1964).

INDEX